Interior Design Review Volume 18

室内设计奥斯卡奖——
第18届安德鲁·马丁国际室内设计大奖获奖作品

（英）马丁·沃勒　编著
安德马丁文化传播　总策划
凤凰空间　出版策划

江苏凤凰科学技术出版社

或许从未有一个行业如室内设计般受女性的影响如此之大。十九世纪时，以男性为主的家具装饰零售商和柜橱制造业主宰着装饰行业。由此导致的便是我们印象中维多利亚时代深沉、硬朗、男性化的室内风格。

1874年，艾格尼丝·加勒特（Agnes Garrett）和她的表姐罗达·加勒特（Rhoda Garrett）成立伦敦首家室内设计公司。她们不仅是彼时最知名的女性室内设计师，而且是最富盛名、口才极佳的女性权益发声体。她们有一个非凡卓绝的家庭。艾格尼丝的大姐是英国第一位女性执照医生。其幼妹米莉森特（Millicent）是全国妇女选举联盟自1897年成立至1918年妇女获得选举权期间，该联盟的主席。

艾格尼丝与罗达合著的《室内装饰中关于绘画、木工和家具的建议》一书首版于1876年，很快登顶畅销书榜。她们倡导自由思考的个人主义、政治行动主义和与之相应的经济独立。

1881年英国女权主义者玛丽·哈维斯（Mary Haweis）出版著作《装潢的艺术》，挑战当时模式化的装潢风气。她声称"我最深的信念，也是作为高尚品位的标准，就是我们的居所理应呈现居住者的个体品位和习性"。

伊迪斯·华顿（Edith Wharton）是亨利·詹姆斯（Henry James），让·高克多（Jean Cocteau）和西奥多·罗斯福（Theodore Roosevelt）的朋友，曾三次入围诺贝尔文学奖。1897年她写就《房屋的装饰》一书，同样表现出对当时设计规则的反抗。

但是，任何人都无法企及艾尔西·德·沃尔夫（Elsie de Wolfe）引发的轰动效应。她于1907年设计完成的麦迪逊大道"殖民俱乐部"，其轻盈灵动的室内装饰彻底重塑了以女性为核心的室内设计。正如她所言"家所体现的永远是女主人的个性，男人只是无数个过客而已"。她非凡职业生涯中的客户名单包括范德比兹家族、摩根家族、罗斯柴尔德家族、弗里克家族和温莎公爵及公爵夫人。

她还曾在第一次世界大战中担任护士，调制出"粉红女郎"鸡尾酒，被称为全球最佳着装女性，甚至在科尔·波特（Cole Porter）的音乐里被永恒传颂。

女性设计师的至高传奇非萨里·毛姆（Syrie Maugham）莫属。她是"儿童之家"创始人贝纳多医生的女儿，企业家戈登·赛弗里奇和威尔士王子的情妇，亨利·维尔康和萨默赛特·毛姆的妻子。

她于1927年设计的全白空间可谓永恒经典。

并非偶然，以上传奇偶像都秉承了独立自主的生活态度。反抗十九世纪重男轻女的社会价值，树立美学观点，参与女性选举权运动，她们为二十世纪中期的女性主义运动奠定了基石。

1915年，伦敦黄页中的127位室内设计师有十位女性，如今67%都是女性设计师。

收录于安德鲁·马丁设计年鉴中的女性设计师们都是她们光彩夺目、勇于开拓的先辈们的后继者。

Martin Waller

Paolo Moschino and Philip Vergeylen
for Nicholas Haslam

Paolo Moschino and Philip Vergeylen. Nicholas Haslam Ltd, London, UK. An award winning interior design company which comprises two retail showrooms in central London, a showroom at the Design Centre in Chelsea Harbour and an international trade office. Recent projects include a chalet in Verbier, a country home in New York and a Kensington apartment. Current work includes three central London townhouses, a private residence in Holland Park, a large country house in Gloucestershire and a Sussex farmhouse. Design philosophy: understand the client and be original.

设计理念：读懂客户，做原创设计。

Eric Kuster

Eric Kuster. Eric Kuster Metropolitan Luxury, The Netherlands. Recent projects include Nikki Beach resort Koh Samui, a luxury penthouse in Monaco as well as the official launch of the Eric Kuster Metropolitan Luxury furniture and textile collection. Current work includes a large residential villa in New York, an extraordinary villa on Ibiza and a reputed hotel restaurant of a 3* Michelin Chef in The Netherlands. Design philosophy: contrast is happiness.

设计理念：快乐对比。

Mary Douglas Drysdale

Mary Douglas Drysdale. Drysdale Design Associates, Washington, DC, USA. Specialising in historic properties for modern life. Current projects include the renovation of a brick country house dating from 1752 and the renovation and decoration of a grand suburban estate built in the 1920's. Recent work includes the design of luxury bathrooms for American Standard Plumbing, the redesign of a New England cow shed into a private dwelling and the renovation of a 1904 Federal style town home in the fashionable Dupont Circle area. Design philosophy: seeing architectural spaces and decoration with a modern lens.

设计理念：以现代三棱镜透视建筑空间和装饰设计。

BRICKS

James van der Velden. BRICKS Amsterdam, The Netherlands. A small practice specialising in residential and commercial projects. Recent work includes the lobby of the citizenM hotel New York, a garage conversion into a loft and a loft on the Amsterdam canal. Current projects include a luxurious chalet in Rougemont, Switzerland, a large family home in Amsterdam and an old school conversion into a family home in the East of Amsterdam. Design philosophy: to contrast old and new.

设计理念：新旧对比。

Allison Paladino

Allison Paladino. Allison Paladino Interior Design, Jupiter, Florida, USA. Specialising in high end, award winning interiors, yachts and commercial design plus luxury home collections. Recent projects include a residence in Palm Beach, Florida, a mansion in Rye, NY and a residence in an exclusive sports club in Jupiter, Florida. Current work includes the main and guest house at a thoroughbred horse farm in Ocala plus multiple houses in the Bahamas and a water front estate in Palm Beach. Design philosophy: timeless design which realises the clients' vision.

设计理念：以经典永恒的设计实现客户的美好愿景。

Design MVW

Ming Xu & Virginie Moriette. Design MVW Co Ltd, China. Specialising in luxury architecture, interior design and furniture design worldwide. Current projects include a three storey office on The Bund, Shanghai, a five star resort hotel in Lijiang, a historic tourism city in Yunnan province and Shanghai Tang stores for China & abroad. Recent work includes Shanghai Tang Cathay Mansion in Shanghai, Giorgetti Furniture Collection and Yun Long Club in Xuzhou. Design philosophy: to balance function and aesthetic using pure lines, lively forms and elegant proportions.

设计理念：运用纯净的线条、生动的形式和优雅的比例，寻求使用功能和美学价值的完美平衡。

Helene Hennie. Christian's & Hennie, Oslo, Norway. Although established in 2007, Helene has been in the industry for over 25 years and is highly sought after. She won the Andrew Martin International Interior Designer of the year in 2007 and has featured in the review for fifteen consecutive years. Projects are exclusive and international.

Current work includes the interior of a private yacht, two large restaurants and a seaside penthouse. Recent projects include a Michelin star restaurant, a country house with stables and a modern but cosy mountain lodge. Design philosophy: to create elegant, comfortable and individual interiors.

设计理念：打造优雅、舒适和个性化的室内空间。

Wan Fuchen

Wan Fuchen. Suzhou FuChen Design Studio, China. Work is high end, with an emphasis on hotels, schools, offices, clubs and villas. Recent projects include Changzhou Detan Hotel, Meizhuishi Exhibition Hall in Suzhou and Baisha Lake private villa in Suzhou. Current work includes Shang Linglong Private Villa in Suzhou, Suzhou Tongli Chinese restaurant and Suzhou Xinguang video centre. Design philosophy: to strive for sustainability.

设计理念：可持续发展。

Studio Hertrich &

Adnet

Marc Hertrich & Nicolas Adnet. Studio Hertrich & Adnet, Paris, France. Projects are worldwide, including hospitality, hotels, spas, restaurants and residential. Recent work includes the interior design of the Sofitel Casablanca Tour Blanche Hotel, Morocco, a contemporary renovation of the legendary '70's Agora Swiss Night Hotel in Lausanne and the luxury redesign of Club Med Guilin, China. Current work includes a contemporary styled luxury boutique hotel in Rabat, sumptuous lounges for a club and sports stadium in Paris and a Relais & Chāteau domain in a vineyard in the South of France. Design philosophy: to dream and create, combining functionality and poetry.

设计理念：去梦想、去创造，将功能与诗意相结合。

Yu Feng

Yu Feng. Deve Build Interior Design Institution, Shenzhen, China. Specialising in the interior design of commercial and public spaces. Current projects include The Oriental Club in Shenzhen, a sales centre in Xinjiang, Urumqi and the company's office space in Shenzhen. Recent work includes the interior of a complex, a high end office building and a commercial plaza in Guangdong plus a club in Hainan. Design philosophy: to preserve traditional Chinese civil engineering whilst embracing modern space design.

设计理念：保持传统城市发展动力，打造现代城市空间设计。

Hare
+
Klein

Meryl Hare. Hare + Klein, Sydney, Australia. An award winning practice specialising in residential and hospitality. Recent work includes an international finance company's headquarters in Sydney, a home on a cliff above iconic Bondi and a home on Sydney Harbour facing the Opera House and Harbour Bridge. Current projects include a luxury motor yacht, a villa in Adelaide and the refurbishment of a 5 star hotel in the Whitsunday Islands in the Great Barrier Reef. Design philosophy: to create original interiors of quality that will stand the test of time and reflect the owner's lifestyle.

设计理念：打造高质量的原创设计，经得起时间的考验，折射客户的生活方式。

Interior Design Philosophy

Jorge Cañete. Interior Design Philosophy, Switzerland. The company's signature is always in search of projects with a poetic dimension. Recent work includes Marie Ducaté's art exhibition in a castle, a mansion in Geneva and a chalet in Chamonix. Current projects include a villa in Capri, a contemporary art museum in Basel and a champagne bar in Paris. Design philosophy: to create personalised projects by analysing three sources of inspiration: the environment in which the project is set, the feeling inherent in the location and the client's own personality.

设计理念：透析三方设计灵感，打造个性化的原创设计——空间所处的环境、置身空间中的心理感受、客户的性格特征。

Chang Ching-Ping

Chang, Ching-Ping. Tien Fun Interior Planning, Taichung, Taiwan. Specialising in high end residential and commercial interiors. Projects are varied including luxury private houses, clubhouses, hotel and office design. Recent work includes two hotels in Taiwan, a yacht club in Shen-Zhen and three show flats in China. Design philosophy: to cultivate fresh ideas.

设计理念：挖掘新知。

75

Sims Hilditch

Emma Sims Hilditch. Sims Hilditch, Gloucestershire, UK. A quintessentially British design practice, specialising in classic yet contemporary interiors for historic buildings as well as luxury residential and commercial properties. Current projects include the conversion of a 16th century pub into a contemporary design studio, a Cotswold manor house and a family mansion in Radlett. Recent work includes a house in Dorset, a Grade I listed crescent development in Bath and several luxury city apartments in London. Design philosophy: preserve authenticity, enhance character.

设计理念：保持权威，彰显个性。

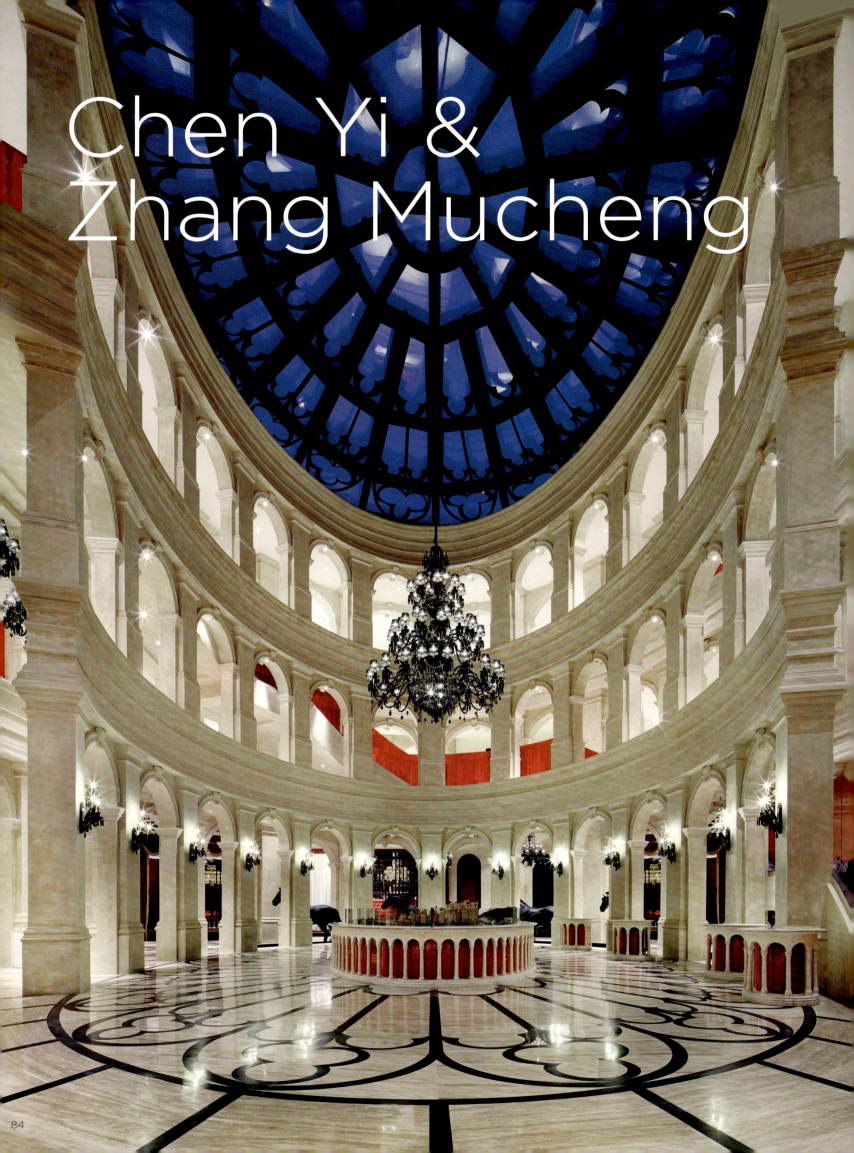

Chen Yi & Zhang Muchen. Beijing Fenghemuchen Space Design Centre, Beijing. Specialising in classic real estate, commercial projects, landscape and product design. Recent work includes a sales centre in Tianjin, a sales club design in Shenyang and another in Xining. Current projects include the architectural and interior design of Blue Lake restaurant and a club in Beijing, Lidu garden landscape design in Baoding industrial park and the showroom design of Tianjin Maple Blue International apartment. Design philosophy: to integrate art and culture.

设计理念：艺术与文化相融合。

Carl Emil Knox

Carl Emil Knox. Carl Emil Knox Design, Binningen, Switzerland. Established in 2005, the practice is increasingly sought after by an international clientele, specialising in building visual identities for offices, hotels, restaurants and public meeting places as well as creating home environments that reflect the values and personality of those who live there. Recent projects include a luxury estate in the South of France, a bachelor's apartment in Stockholm and a family home in London. Current work includes the offices for Scandinavia's leading model agency MIKAs in Malmö and Stockholm, a private holiday home in Southern Italy and a private residence in Basel, Switzerland. Design philosophy: art, simplicity and style.

设计理念：艺术、简约与风格。

Ben Wu

Ben Wu. W. Architectural Design Co, Shanghai, China. Specialising in hotels, clubs and private residences since 1993. Recent projects include Jiangshan 99 private villa in Ningbo and Beijing Cofco Auspicious Palace. Current work includes Sancha Lake Shimao centre sales office in Chengdu and Ningbo Fortune centre residential office apartment. Design philosophy: dedicated to forming a new design language with oriental philosophy and macro-aesthetic trend.

设计理念：以东方哲学思想和宏观美学潮流形成现代设计语汇。

Stefano Dorata

Stefano Dorata. Studio Dorata, Rome, Italy. An architectural practice specialising in apartments, villas, yachts and hotels in Europe, America and Asia. Stefano recently published his own book 'Case' (Houses) by Stefano Dorata, showcasing simplicity and order as the aesthetic hallmarks of his design. Recent projects include a villa on the sand dunes in Sabaudia, an apartment in Rome and a villa in Porto Rotondo, Sardinia. Current work includes four villas in Indonesia, a penthouse in Milan next to Via Montenapoleone and a boutique hotel in the centre of Tel Aviv.

Yu-Lin Shin

Yu-Lin Shin. Dumas Interior Design Group, Taiwan. Specialising in interior design, space planning and project management for residential, commercial and corporate office projects. Recent work includes Mimi's Steak House in Taichung, The Clouds Villas public space in Taichung and The Grand Palace in Taipei. Design philosophy: interior design should not only satisfy our sense of beauty, it should also portray one's attitude to life.

设计理念：室内空间设计不仅可以满足审美需求，还能够彰显生活态度。

Ligia Casanova

Lígia Casanova. Atelier Ligia Casanova, Lisbon, Portugal. Specialising in residential and public spaces internationaly. Recent projects include the interior design for a luxury apartment complex aimed at the short-term rental market in Lisbon. Current work includes the development of a new high end apartment project, a Tribeca loft in New York and a penthouse in São Paulo. Design philosophy: make room for happiness.

设计理念：营造快乐空间。

Nicky Dobree

Nicky Dobree. Nicky Dobree Interior Design. Specialising in luxury ski chalets and high end residential projects internationally. Recent work includes chalets in Gstaad, Klosters, St Moritz, a country house in Surrey and a London apartment. Current projects include chalets in Saas Fee, Val-d'Isère, Morzine and Verbier and a listed house in Windsor. Design philosophy: to create comfortable and beautiful homes which complement the personality and lifestyle of each client.

设计理念：打造将客户的性格特征和生活方式完美结合的舒适、美丽的家。

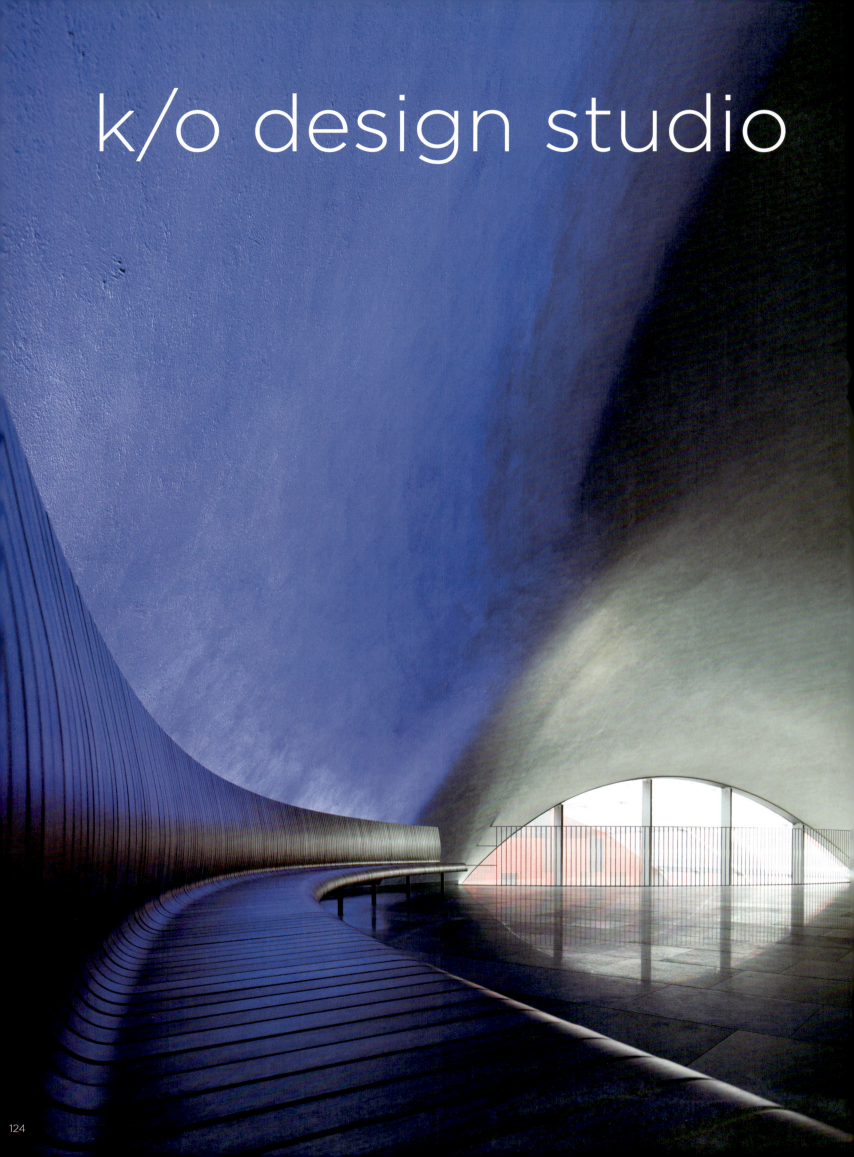

k/o design studio

Kunihide Oshinomi. k/o design studio, Tokyo. Covering residential interiors to skyscraper design in collaboration with professionals around the world. Current projects include a luxury duplex condominium interior next to the National Palace Museum in Taipei and a high rise penthouse interior in the centre of Tokyo. Recent work includes a campus project at a music college and several condominium interiors all in Tokyo. Design philosophy: to provide an amazing environment, through a wide range of design skills including landscape, architecture, interior and furniture.

设计理念：运用一系列广泛的设计技巧，如景观、建筑、室内、家具等，打造充满惊喜的空间环境。

Aleksandra Laska

Aleksandra Laska. Aleksandra Laska, Warsaw, Poland. An avid art collector, specialising in creating individual and timeless interiors. Recent work includes the partial remodeling of the National Opera House in Warsaw, a 1,700 sq m concrete showroom in Warsaw's city centre and a period apartment in Warsaw's old town. Current work includes a 130 sq m showroom located on Plac Dabrowskiego in Central Warsaw, a 70 sq m historic apartment and adjoining studio of painter Piotr Trzebinski and the renovation of a 1902 factory space in Moscow into a theatre. Design philosophy: to successfully combine contrasting elements.

设计理念：将各种对比鲜明的设计元素成功结合。

Maria Barros

Maria Barros. Maria Barros Home, Cascais, Portugal. With a strong Palm Beach aesthetic, bold use of pattern and colour, Maria's influence stems from her time travelling and living in Florida as a teenager. Recent projects include a chic Lisbon family apartment, a fancy beach bar in the Lisbon coastal area and a holiday home in the Algarve. Current work includes redecorating a popular sushi restaurant in Cascais, a fashion boutique in Madrid and a family home in Alentejo. Design philosophy: making places that elevate the spirit. Promoting happiness through design.

设计理念：打造提升精神品位的空间，边设计，边快乐。

Beijing Newsdays

Jianguo Liang, Wenqi Cai, Yiqun Wu, Junye Song, Zhenhua Luo, Chunkai Nie, Eryong Wang. Beijing Newsdays Architectural Design Co Ltd, China. High end commissions including hotels, clubhouses, restaurants, showrooms and public spaces. Recent work includes a renovation project for Wanliu Academy in Beijing, the Turandot Hotel and museums in Liling, Hunan Province and The Emperor Beijing, boutique hotel in Mudu. Design philosophy: serve the client with elegant solutions.

设计理念：以优雅的设计回报客户。

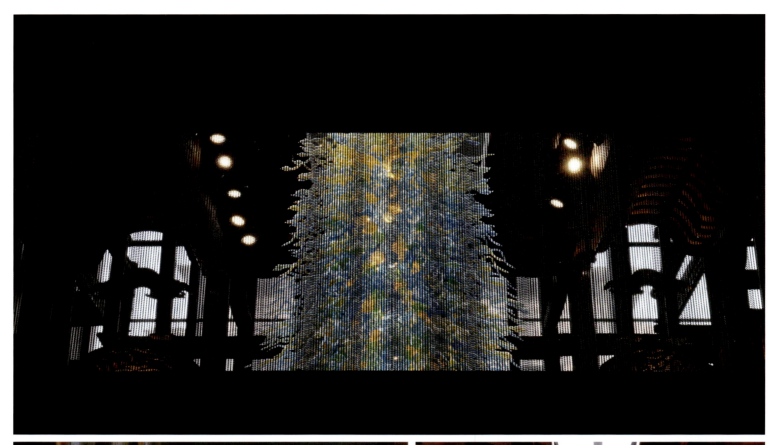

Honky

Christopher Dezille. Honky, London, UK. Specialising in the transformation of high profile residential developments across the Capital. Recent work includes four lateral apartments and a duplex penthouse in Trafalgar Square, a lateral apartment & triplex penthouse with unrivalled 360 degree views in Westminster and an apartment by the water's edge with views of the Tower of London & Tower Bridge. Current projects include a penthouse on the Côte d'Azur, a seven bed family home in Hampstead and a comfortable Cotswold retreat created from agricultural buildings. Design philosophy: innovation, quality & service.

设计理念：创新、品质和服务。

POCO Designs

Charlotte and Poppy O'Neil. POCO Designs, Sydney, Australia. POCO thrive on creating new and innovative environments. Recent projects include the refurbishment of a Sydney waterfront boathouse, the styling of a contemporary Sydney family home and the redevelopment of a waterfront penthouse overlooking Sydney Harbour. Current projects include the design of a modern new office space in Sydney's CBD, the renovation of a waterfront in Sydney's North Shore and a rooftop penthouse with garden overlooking Sydney Harbour Bridge. Design philosophy: bring to life the style, passion and imagination of the client.

设计理念：给客户的生活带来时尚、热情和想象。

Hong Zhongxuan

Hong Zhongxuan. HHD Eastern Holiday International Design, Shenzhen, China. Pioneers in luxury and boutique hotels & resorts in China. Current projects include Modern Classic Hotel, Renhe Spring Hotel and Tianjin Light Hegu resort & spa. Recent work includes Beauty Crown 7 star Hotel in Sanya, Crowne Plaza Hotel Baoji and The Liuzhou Hotel in Shanghai. Design philosophy: functional, comfortable and elegant.

设计理念：功能、舒适和高雅。

Angelos Angelopoulos

Angelos Angelopoulos. Angelos Angelopoulos Associates, Athens, Greece. Work is international, specialising in boutique hotels, private residences, apartments, restaurants, clubs, hotels, mountain & sea resorts, showrooms, workspaces, fabric design and conceptual design. Recent projects include a new beach resort in Cyprus, the architectural and interior design of a private estate in Athens and a private summer residence on a Greek island. Current work includes a restaurant in New York State, exclusive VIP villas at a beach resort in Cyprus and a private residence by the sea in Attica, Greece. Design philosophy: psychology and self expression.

设计理念：心理学和自我表达。

Idmen Liu

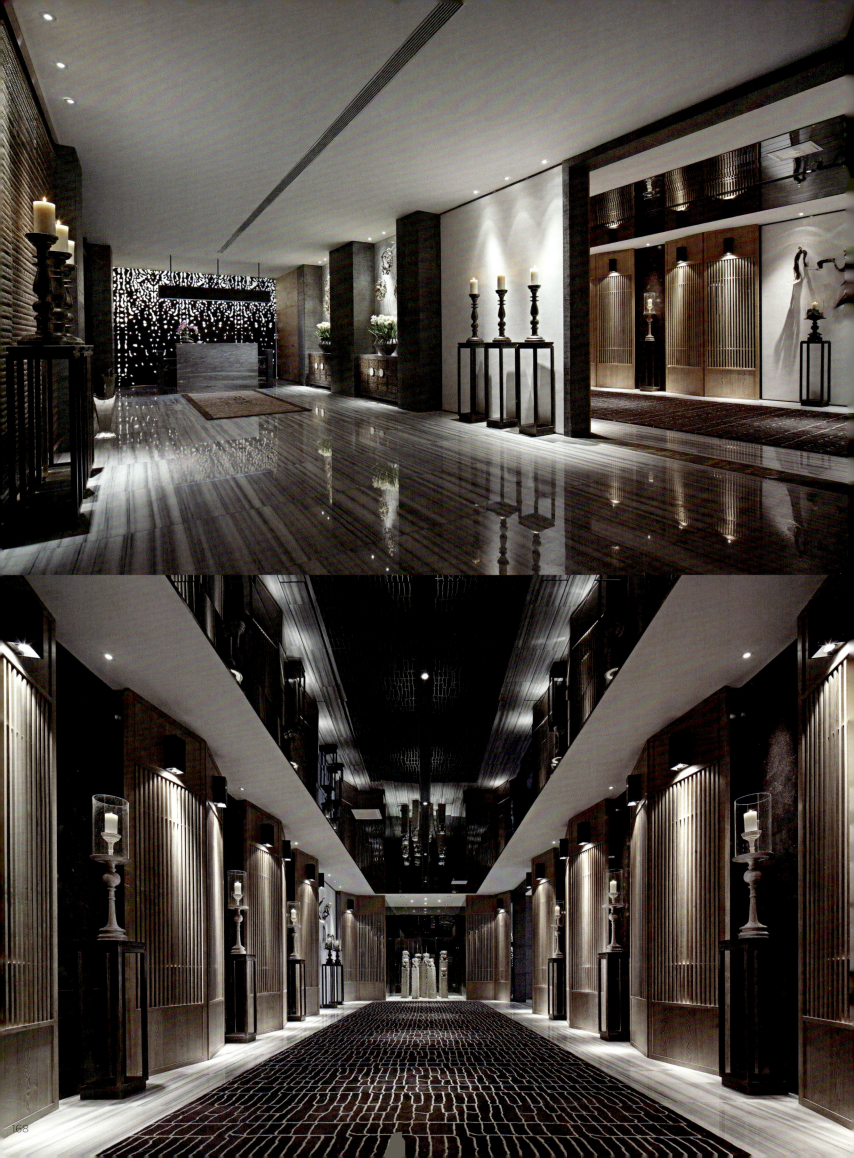

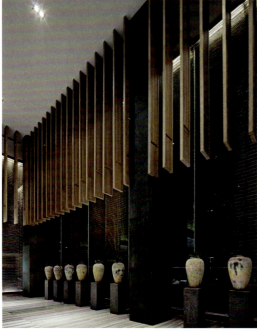

Idmen Liu. Shenzhen Juzhen Mingcui Design Co Ltd, Shenzhen, China. Dedicated to high-class and deluxe style interior design. Recent projects include Vanke Cheerful Bay Club House in Chongqing, Vanke City sales centre in Zhengzhou and Vanke Chengdu District office in Chengdu. Current projects include Wonders Ziyuntai Club House in Quanzhou, Tuo Ji Hong Bao Company office in Shenzhen and Vanke Jin Yu Xue Fu sales centre in Chongqing. Design philosophy: keep walking on the road of design.

设计理念：在设计的道路上勇往直前。

Holly Stone & Nashmita Rajiah. Intarya, London. An award winning interior design & architecture studio creating luxury residential interiors in London and around the world. Current projects include a substantial property in Belgravia and a prime residential development in St James. Recent work includes a Grade II listed house in Kensington and several apartments in Chelsea & Hyde Park. Design philosophy: to emphasise original features with a sympathetic approach.

设计理念：通过讨巧的手法来强化原始特色。

Elin Fossland

Elin Fossland. ARKITEKTFOSSLAND AS, Drammen, Norway. Specialising predominantly in private residences with some public space including schools, restaurants and hotels. Elin is renowned for combining modern design with antique furniture to create exciting interiors. Recent projects include a summerhouse by the sea, a family house in Drammen and a wine bar in the mountains. Design philosophy: to create functional environments that reflect the client's personality.

设计理念：打造反映客户个性的功能环境。

SAARANHA&VASCONCELOS

Rosário Tello, Carmo Aranha. SAARANHA&VASCONCELOS, Lisbon, Portugal. Specialising in predominantly residential projects with some commercial, including private yachts. Recent work in Lisbon includes a large family house on the outskirts, a small bachelor apartment in the centre and the So Chic concept store in downtown Chiado. Current projects include the interior design for several houses within a large rural private hunting property in Alentejo, Portugal, a family home outside Lisbon and the renovation of an historic building in the old town. Design philosophy: to meet client expectation with originality.

设计理念：满足客户对创新性的期望。

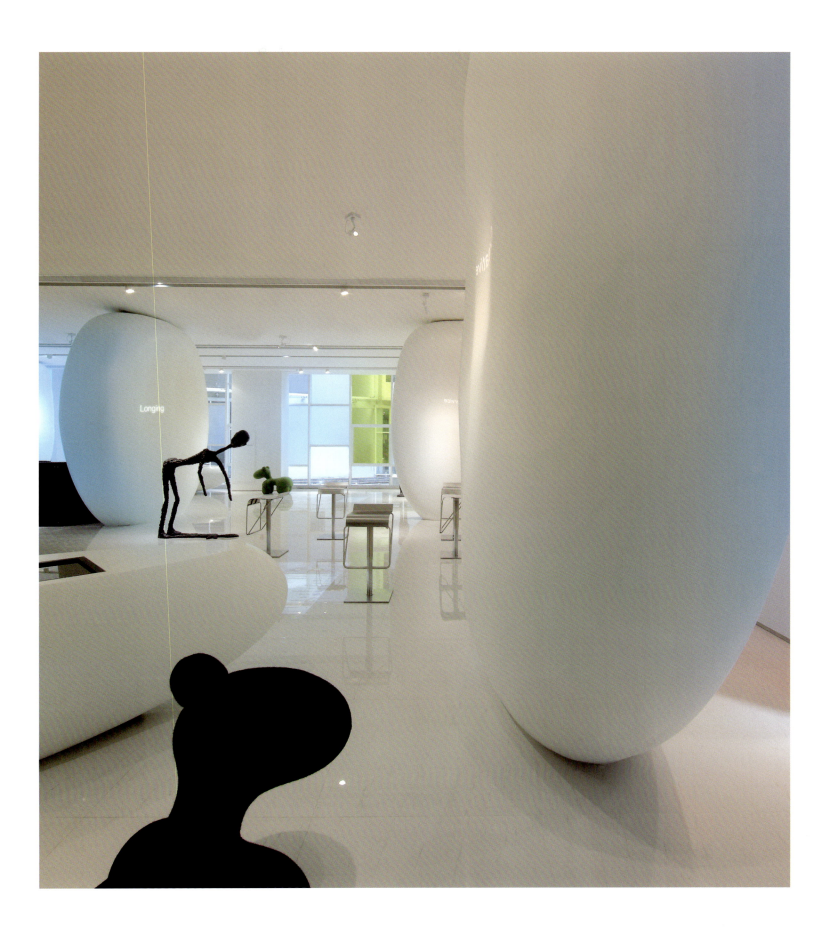

Gu Teng

Gu Teng. Times Property Holdings Limited, Guangzhou, China. Specialising in real estate investment, development and management. Recent projects include Huangsheng sales centre in Guangzhou, Shidainanwan sales centre in Guangzhou and Shidaiqingcheng sales centre in Zhuhai. Current work includes Yunduan sales centre in Zhuhai, Times Experience Centre in Guangzhou and Xilong Living Club in Qingyuan. Design philosophy: to incorporate art with life.

设计理念：艺术与生活相融合。

AZULTIERRA

Toni Espuch. AZULTIERRA, Barcelona, Spain. Specialising in luxury residential and commercial interiors around the world. Current projects include a restaurant in Barcelona, the showroom of a major natural products brand and a luxury family house. Recent work includes Black Cocktail Bar, a high end home in Barcelona and a shop for a famous shoe brand. Design philosophy: to create warm, atmospheric spaces with character.

设计理念：打造具有温馨氛围和个性的空间。

Maurizio Pellizzoni. MPD London, UK. An award winning practice offering a tailored interior design service with a portfolio that includes prime residences in the UK, the US and Europe. Recent work includes a Shoreditch loft, Ascot Lodge, a Grade II listed building dating from the 15th century used by King James as his hunting lodge and the full structural redesign and build of Antigua House, Chelsea. Current work includes the extension of a large Edwardian property, the redesign of a large family house on the edge of the Willamette River in Portland, Oregon and a two bedroom lateral apartment in a prestigious new Berkeley development in Belgravia. Design philosophy: to combine neo-classical luxury with quality craftsmanship and innovative design solutions.

设计理念：将新古典奢华与高品质工艺和创新设计方案相结合。

Katharine Pooley

Katharine Pooley. Katharine Pooley, London & Qatar. Specialising in creating classic, contemporary and high end landmark commissions worldwide. Work includes the recently opened 3,000 sq ft Katharine Pooley showroom in Doha, the bespoke design of a Grosvenor Square apartment and the sensitive redesign of a 17th century Cotswold cottage. Current projects include a

cutting edge, fifty bedroom boutique hotel development with spa in China, the contemporary design of a 10,000 sq ft private villa with cinema and gym in Doha and the redesign of a listed 10,000 sq ft Mayfair townhouse. Design philosophy: to combine contemporary living with flexibility, context and craftsmanship with an unrivalled attention to detail.

设计理念：将现代生活与灵活性、环境和无可挑剔的细节工艺结合在一起。

Bronnie Masefau

Bronnie Masefau. Bronnie Masefau Design Consultancy, Victoria, Australia. A multidisciplinary design consultancy. Current projects include a large family retreat on the Morning Peninsula, a commercial auditorium with lobby, café and shop in Queensland and a six star dwelling in the heart of Melbourne. Recent work includes an award winning beach house on the Great Ocean Road, a café and salon in Chiang Rai, Thailand and an Art Déco styled multi-level family home in Melbourne. Design philosophy: to design homes and workplaces with purpose, function and soul.

设计理念：使设计的住宅及办公项目集用途、功能和内涵于一身。

Ana Cordeiro

Ana Cordeiro. Prego Sem Estopa, Lisbon, Portugal. Recent work includes a hotel in the south of Portugal, a coffee and gourmet shop in Lisbon and the company's own furniture range. Current projects include several residential and commercial interiors in Portugal and a family house with garden in the centre of London. Design philosophy: elegant and comfortable interiors for a happy life.

设计理念：为幸福的生活打造典雅、舒适的室内设计。

Plus Design

Prasetio Budhi. Plus Design, Jakarta, Indonesia. Specialising in newly built, high end residences and commercial spaces. Current projects include a large family home for a well known entrepreneur and a mansion complete with ballroom and Olympic pool. Recent work includes a retail boutique in Bangkok and resort villas in Bali. Design philosophy: simple elegance.

设计理念：简约典雅。

Intimate Living Interiors

Kari Arendsen & Jenn Bibay. Intimate Living Interiors, California, USA. An Internationally renowned firm specialising in high end residential design. Recent work includes two 4000 sq ft ranch houses in Rancho Santa Fe, California, the guest suite of a Pasadena show house and a 1200 sq ft

Philadelphia loft at the Ritz Carlton. Current projects include the renovation of a 3500 sq ft Colonial revival residence for a celebrity chef, a 4000 sq ft modern Mediterranean house in Encinitas, California and a 4500 sq ft East Coast home for a professional athlete in Naples, Florida. Design philosophy: blending the reclaimed with the refined and the unexpected with the essential.

设计理念：将质朴与精致、惊喜与平凡相融合。

Candy & Candy

Nick Candy. Candy & Candy, London, UK. Founded by Nick and Christian Candy in 1999, Candy & Candy is one of the world's leading interior design houses dedicated to designing the most luxurious real estate. Current projects include a Kensington penthouse, two properties in Cadogan Square, Knightsbridge and two further residences in Eaton Square. Recent work includes a five bedroom duplex penthouse with 360 degree views in Arlington Street, London, directly above The Wolseley restaurant and Cheval House, Knightsbridge a contemporary Art Déco styled five bedroom duplex penthouse. Design philosophy: driven by the pursuit of perfection.

设计理念：不断追求完美。

Laura Brucco

Laura Brucco. Laura Brucco, Buenos Aires, Argentina. An interior architecture and design studio specialising in luxury residential, commercial and corporate interiors. Current projects include a villa in the most exclusive area in Buenos Aires as well as several large family houses. Recent work includes the interior architecture of a villa on the outskirts of the city, an impressive duplex for a tennis star and several flats in high end towers in the residential district. Design philosophy: timeless elegance.

设计理念：永恒的典雅。

Joseph Sy

Joseph Sy. Joseph Sy & Associates, Hong Kong. Established in 1988, Sy has a strong passion for creative lighting and space planning and is often invited to lecture in Hong Kong and Mainland China. Current projects include clubhouses in Chongqing and Chengdu, an office project in Shenyang and lighting showrooms in various cities in China. Recent work includes a lighting showroom in Kunshan, a four storey villa in Shenzhen and a jewellery shop in Hong Kong. Design philosophy: effective and functional space planning is vital.

设计理念：空间的效用及功能规划至上。

Alejandro Niz & Patricio Chauvet

Alejandro Niz & Patricio Chauvet. Niz + Chauvet Arquitectos, Mexico City. Established in 2002, specialising in a wide range of residential projects, restaurants and hotels. Recent work includes a private family home in Reforma Lomas, a hotel & spa in Arcos Bosques complex and a set of restaurants in the Santa Fe shopping centre, Mexico City. Current work includes projects for hotel developers, including Hotel Thompson in Playa del Carmen and the national hotel group Posadas. Design philosophy: functionality and comfort.

设计理念：功能与舒适。

Tim Campbell. Studio Tim Campbell, Los Angeles, California, USA. A boutique practice specialising in high end custom residential design for renovations and new construction as well as commercial projects including galleries, restaurants and hotels. With offices in Los Angeles and New York City, the studio has completed projects in the United States, Mexico and abroad. Recent work includes one of the newly renovated loft style interiors in the iconic 19th century Printing House, New York, NY, a modern retreat in Silverlake, Los Angeles to house Tim's own contemporary art collection and the sensitive restoration of Richard Neutra's Singleton house in Beverly Hills. Design philosophy: to create individual spaces which reflect each client's unique sensibility.

设计理念：打造反映客户独特品位的个人空间。

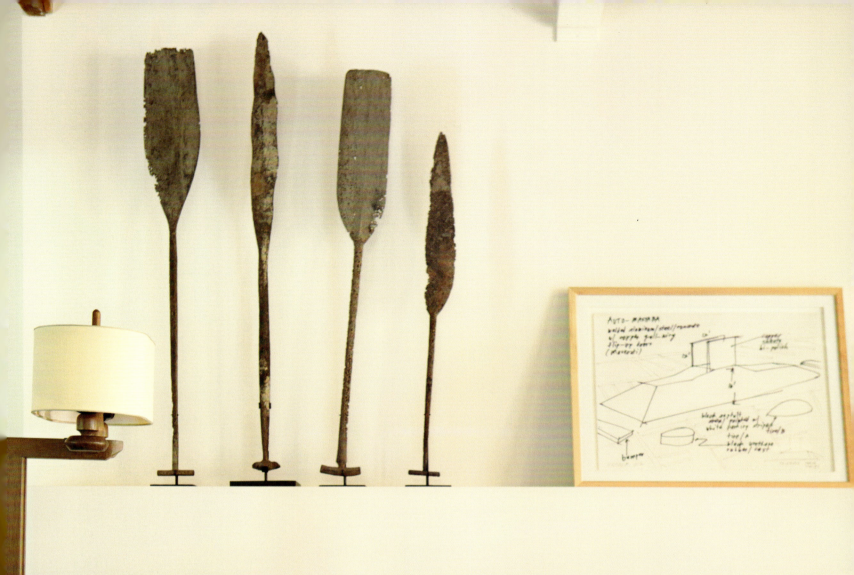

Tim Campbell

Ben & Hamish Lewis. TRENZSEATER, Christchurch, New Zealand. Specialising in high end luxury residential and commercial interior design both in New Zealand and internationally. Current work includes a French Chateau inspired country estate, an established grand home in Christchurch and several comprehensive large scale private residences around New Zealand. Recent projects include prestigious apartments in Auckland, Christchurch and Queenstown plus several high end private residences and holiday homes. Design philosophy: elegant and balanced.

设计理念：典雅与平衡。

Olga Stupenko

Olga Stupenko. Olga Stupenko Design, Oxfordshire UK & Moscow, Russia. Specialising in high end residential and commercial interior and architecture globally, with offices also in London and Monaco. Recent projects include several private mansions in Moscow. Current work includes a luxury hotel in France, a movie themed apartment in Moscow and a penthouse in Monaco. Design philosophy: to use simple shapes and proportions to create unique and timeless interiors.

设计理念：用简单的形状和比例打造独特而永恒的室内空间。

Jiang Jianyu

Jiang Jianyu. Hangzhou Daxiang Art Design Co, Hangzhou, China. Specialising in commercial work. Current projects include an International brand hotel and several Chinese restaurants and spa projects. Recent work includes the restaurant of InterContinental Beijing Financial Street and Shangyu International Hotel, Zhejiang province. Design philosophy: individual design combined with local tradition.

设计理念：个性化设计与地方传统相结合。

Platform 9

Elaine McBride & Gillian Monro. Platform 9 International, Glasgow, Scotland. Offering an exclusive service for property developers, private clients and hoteliers throughout the UK as well as internationally. Current projects include luxury homes in Florida and commercial hotel projects. Recent work includes the interiors of private motor yachts, boutique ski chalets in France and luxury lodges. Design philosophy: to deliver unique environments which exceed expectation.

设计理念：实现超出预期的独特环境。

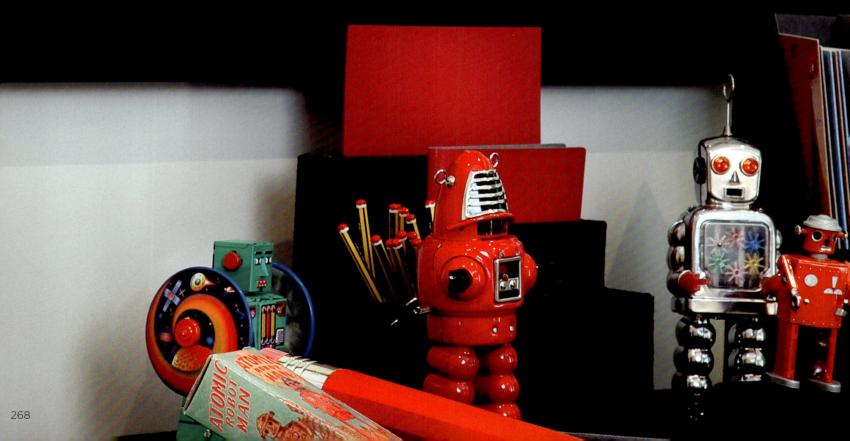

Alexander James

Stacey Sibley. Alexander James Interior Design, Berkshire, UK. Delivering a complete service for luxury developers and high end residential clients. Recent projects include a landmark contemporary house in Surrey with views across three counties, apartments in Mayfair and Marbella and luxury hotels across Eastern Europe. Current work includes a residential refurbishment in Barbados, a large riverside house and a distinctive family home on the Wentworth Estate. Design philosophy: excellence and detail.

设计理念：完美与细节。

Detlev Böhnke

Detlev Böhnke. Paris 56 - fine interiors, Berlin, Germany. Specialising in residential interiors from small to high end luxury. Recent work includes a penthouse in the South of France, apartments in Berlin, Monaco and Moscow and a villa in the countryside near Moscow. Design philosophy: originality and attention to detail.

设计理念：追求创新、注重细节。

Suzanne Lovell

Suzanne Lovell, Suzanne Lovell Inc. Chicago, Illinois, USA. Luxury residential interior architecture, design and fine art. Recent projects include a private owner's suite at the St. Regis, Manhattan, a landmark Howard Van Doren Shaw penthouse on Chicago's Lakefront, a San Francisco Penthouse and Carmel Beachfront Estate. Current work includes an oceanfront penthouse in Miami, a villa in the South of France and a family estate in The Hague. Design philosophy: to create couture environments for extraordinary living.

设计理念：为非凡的生活打造定制的环境。

Bo Li

Bo Li. Cimax Design Engineering, Shenzhen, China. A young team providing a variety of services including interior, architectural, exterior and product design. Recent projects include the renovation of a lakeside club, a private mansion and a villa show flat near the Lijiang River. Current work includes the landscape planning and interior for a high end villa community, the interior design for a large entertainment and sports club and the architectural and interior design for a seaside resort hotel. Design philosophy: focus on the user's experience.

设计理念：注重使用者的体验。

Powell & Bonnell

David Powell & Fenwick Bonnell. Powell & Bonnell, Ontario, Canada. Specialising in the creation of international luxury residences, as well as an eponymous collection of bespoke furniture, lighting and textiles. Recent projects include transforming an abandoned stone ruin in the Shetland Islands into an isolated contemporary getaway, updating a sprawling Hampton's estate and converting an 18th century Toronto church into a high-style family residence. Current work consists of residences located in New York, Chicago, Palm Beach and Toronto. Design philosophy: to provide tailored, lifestyle-centric design solutions of uncompromising quality.

设计理念：提供高品质的定制、以生活方式为中心的设计方案。

287

289

Pippa Paton

Pippa Paton. Pippa Paton Design, Oxfordshire, UK. Specialising in the complete renovation, remodelling and interior design of residential projects in the Cotswolds. Recent work includes a 12,000 square foot Victorian residence in Oxfordshire, the renovation of a Georgian rectory and the transformation of two listed barns to create a contemporary home. Current projects include a small Gloucestershire estate, a listed manor house and Cotswold cottage and a new 'legacy' home. Design philosophy: to enhance the enjoyment of living in your home.

设计理念：强化居家生活的乐趣。

Tomoko Ikegai

Tomoko Ikegai. ikg inc, Tokyo, Japan. Established in 2006, ikg undertake a wide range of architectural and interior design services as well as the selection and coordination of art, furniture, electrical and household items. Recent projects include a poolside house for a family in Tokyo, a solar powered eco house and Matsukura Hebe Daikanyama clinic. Current work includes a commercial complex in Tokyo, a holiday home by the beach and the renovation of various residences. Design philosophy: to provide an environment to inspire and invigorate.

设计理念：创建极具灵感、鼓舞人心的环境。

Julia Buckingham

Julia Buckingham. Buckingham Interiors + Design, Chicago, Illinois, USA. Current projects include a 16,000 square foot new build home on Lake Michigan, the fusion of 2 luxury penthouse condominiums in Phoenix and a complete redesign of a historic George Maher landmark in Chicago's Hyde Park. Recent work includes the renovation of a classic greystone townhouse in Chicago's Lincoln Park, the interior design of a 12,000 square foot custom built home in Chicago's North Shore and the restoration of a grand Victorian manse in Iowa. Design philosophy: blending old and new to reveal the beauty in unexpected pairings, a term which Julia coined Modernique®.

设计理念：如朱莉杜撰的词汇"Modernique"，旨在通过意想不到的搭配使新与旧融合在一起。

Ann Yu

Ann Yu. Guangzhou DOMANI Decoration Design, China. Specialising in a wide range of space design and corporate planning in the Pacific region. Recent projects include Foshan Luochun Timesgroup sales club, ZhongHai Hefei villa and Rongsense Shimao office design. Current work includes KunMing Huaxia Rerl club, NanChang Zhenro Group runcheng sales club and Huidong TanYue bi-monthly bay hotel design. Design philosophy: dedicated to excellence.

设计理念：追求完美。

Karen Akers

Karen Akers. Designed by Karen Akers, NSW Australia. A boutique practice specialising in residential design. Current projects include a heritage listed, Gothic style residence in inner Sydney, a beach front holiday house on the Central Coast of New South Wales and a heritage listed residence on Sydney harbor. Design philosophy: tailor made, combining new and old elements.

设计理念：专属定制，将新旧元素相结合。

Associates III

Kari Foster and design team Angie Pache, Jill Bosshart, Renee Keller, Rachel Blackburn, Michaela Jenkins, Jason Schleisman. Associates III Interior Design, Denver, Colorado, USA. Innovative responsible designers creating beautiful, healthy, and nurturing interiors for eco-conscious clients worldwide. Current projects include a modern family home in Vail, an Aspen Valley ranch and a flat in Houston. Recent work includes a California beach home, Colorado mountain retreat and a downtown Los Angeles hi-rise apartment, each a contemporary residence in which the interiors were a backdrop to the views. Design philosophy: inspiring change through environmental awareness.

设计理念：鼓舞人心的改变源于环保意识。

ShuHeng, Huang. Sherwood Design, Taipei, Taiwan. An award winning internationally known design company specialising in bespoke work. Recent projects include mansions in Sham Chun, two sales centres in Hebei Province as well as architecture in Chengdu and four club houses in Taiwan. Design philosophy: dedicated to creating a balance between classical and modern, eastern and western, artificial and natural.

设计理念：致力于打造集古典及现代，东方与西方和人工与自然之间的平衡。

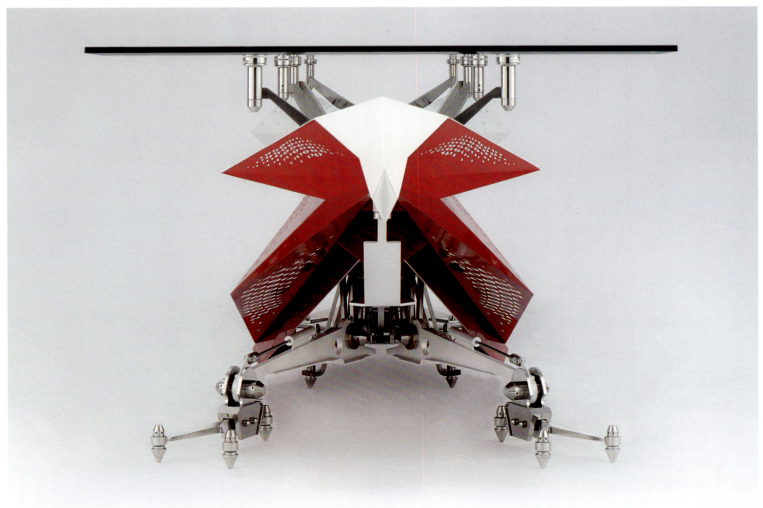

Jennifer Mehditash

Jennifer Mehditash. Mehditash Design, California & New York, USA. A boutique practice with international projects. Current work includes a home in Martha's Vineyard and a large country home in Darien, Connecticut. Recent projects include the remodel and design of a room for a children's charity in the Ronald McDonald House, Long Island, apartments throughout Manhattan, large scale residences in the Hamptons, Connecticut and Westchester, NY plus various private residences, hotels and spas throughout Portugal. Design philosophy: to create beautiful, tailor made interiors.

设计理念：打造完美、专属的室内空间。

Taylor Howes

Karen Howes. Taylor Howes Designs, London. An international interior design firm who have carved a niche in offering a luxury comprehensive design service to private clients, property developers & hoteliers. Current projects include a glamorous apartment in Grosvenor Square, a six storey townhouse in Kensington and an old rectory in Devon. Recent work includes a marketing suite for a well known developer on Berkeley Square, two show apartments in Mayfair's Jermyn Street and a sophisticated penthouse apartment in Chelsea. Design philosophy: to maintain creative excellence and friendly service.

设计理念：保持卓越的创造能力及良好的服务。

Gloria Cortina

Gloria Cortina, Vanessa Ocaña, Rafael Franco. Gloria Cortina Estudio, Mexico. A leading design studio specialising in city, country and beach homes to luxury retreats all over the world. Recent projects include a contemporary high-end residence set in a 4000 sq meter landscaped plot, a Mexico city apartment with panoramic views and Rancho LB, a multi acre country retreat. Current work includes Las Lomas, a home tailor made to display a collection of books and eclectic art pieces, Cabo 136 a beach house in Baja California, which showcases the company's first collection of beach furniture combined with Italian and Mexican fabrics. Design philosophy: signature concepts with the personality and textures of modern day Mexico.

设计理念：打造个性与现代墨西哥风格的理念。

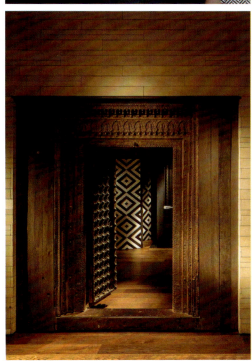

Steve Leung

Steve Leung. Steve Leung Designers, Hong Kong. Consisting of a team of 350 talented staff who specialise in residential, commercial and hospitality design. Recent projects include Eden Residences Capitol in Singapore, Han Yue Lou Villa Resort in Huangshan and Yuan at Atlantis The Palm in Dubai. Current work includes a hotel in London, a serviced apartment project in Ulaanbaatar and a resort hotel in Sanya, China. Design philosophy: Enjoy Life · Enjoy Design.

设计理念：享受生活，享受设计。

CarterTyberghein

Patrick Tyberghein. CarterTyberghein, London, UK. Specialising in luxury interior architecture in the UK and overseas including both primary and secondary private homes, residential developments and boutique hotels. Current projects include a villa in the South of France, a substantial seaside house in Jersey, major homes in Surrey and executive offices in Scotland. Recent work includes apartments in Hampstead, Regents Park and Kensington, a beach house in Barbados and a large family home in Wimbledon. Design philosophy: tailor made elegance.

设计理念：高端定制。

One Plus Partnership

Ajax Law Ling Kit & Virginia Lung Wai Ki. One Plus Partnership Ltd, Hong Kong. Established in 2004, by 2012 One Plus were the first Asian interior design firm to win the Andrew Martin International Interior Designer of the Year Award. Their creative inspiration for each project is focused on a specific theme.

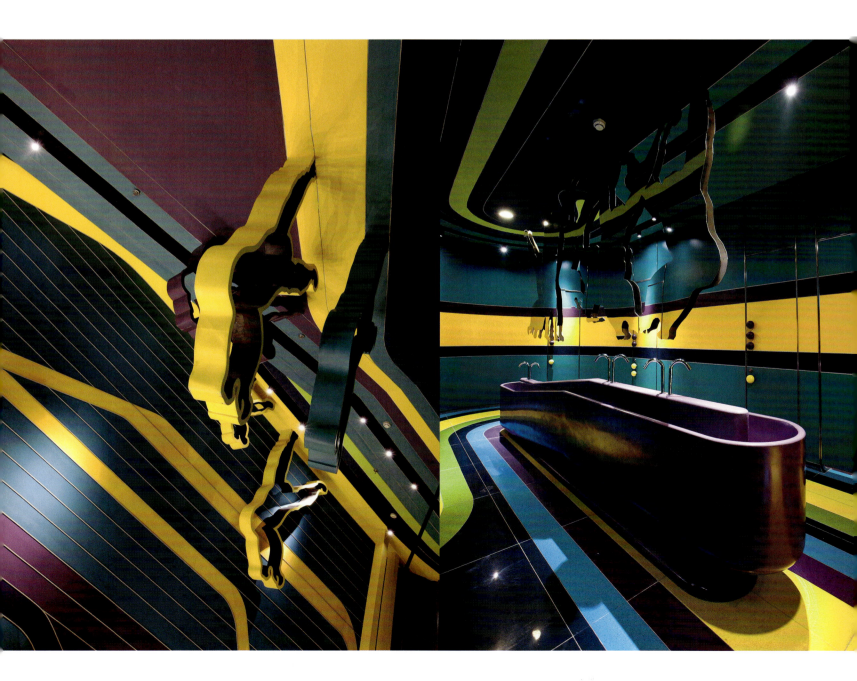

Current work includes the Cine Times, Nanchang Insun International Cinema and One Plus Partnership Office. Recent projects include the interior of several cinemas and sales office projects in China and a club house in Hong Kong. Design philosophy: bold and diverse.

设计理念：大胆与多样化。

Mona Hajj

Mona Hajj. Mona Hajj Interiors, Baltimore, USA. Focusing on high end residential and commercial projects globally. Recent work includes a waterfront condo in San Francisco, an interior and architectural renovation in Georgetown, DC and a horse farm in Kentucky. Current projects include a Beverly Hills villa, a period renovation in the historic Guilford area Baltimore and a farmhouse in Connecticut. Design philosophy: combine a global vision with an American emphasis on elegance, comfort and simplicity.

设计理念：将美式中高端、舒适与简易的风格与国际视野相融合。

Oleg Klodt

Oleg Klodt & Anna Agapova. Architecture and Design by Oleg Klodt, Moscow, Russia. Bespoke interiors for residences and apartments in the luxury sector as well as projects for public buildings. Recent work includes an impressive home near Rublevskaya Highway, inspired by Soviet era Constructivism, a large Art Déco style apartment and a retail outlet for the French clothes company 'Loft design by...' Current projects include numerous large family homes in Moscow, a new meat themed restaurant and a range of Oleg Klodt designer furniture. Design philosophy: understated, harmonious, eclectic.

设计理念：低调、和谐、不拘一格。

Jin Jian

Jin Jian. ZPSS. (Zhu Ping Shang She.) Hangzhou, China. Founded in 2009, ZPSS specialise in interior and furniture design. Recent projects include Zhupin coffee shop in Hangzhou, Guqiao restaurant in Chongqing and Shangshangting private villa in Hangzhou. Current work includes Shanghai boutique hotel, Moganshan resort hotel and Yangzhai private house in Nanjing. Design philosophy: art and design is for life.

设计理念：艺术与设计皆为生活。

Broosk Saib

Broosk Saib. Broosk Saib, London. Specialising in high end residential interiors in the UK and abroad. Recent projects include a family house in Wimbledon, a bachelor pad in central London and a large apartment in Stockholm. Current work includes a large family house in Belgravia, a villa in Riyadh and an apartment in Chelsea. Design philosophy: to combine Eastern splendor with Western luxury.

设计理念：将东方的壮丽与西方的奢华相融合。

The Studio Harrods

Sheena Notley-Griffiths, Claudia Fuchsberger, Adriel Lack & Henry Prideaux. The Studio Harrods, Knightsbridge, London, UK. A bespoke service offering original design from concept through to installation on a range of residential and commercial briefs. Recent projects include a pair of penthouse apartments in Chelsea, a contemporary family home in Richmond and a private residence in South Kensington. Current work includes a luxury villa in the Middle East, an exclusive lateral apartment in Knightsbridge and a distinctive mews house in Mayfair. Design philosophy: tailor made luxury for an international clientele.

设计理念:为各国客人奢华定制。

Kathleen Hay

Kathleen Hay. Kathleen Hay Designs, Nantucket MA, USA. With an emphasis and expertise on new construction for the luxury residential and commercial market. Current projects include the redesign of a modern apartment in an old mill in Brooklyn, a pied-à-tèrre in Palm Beach and a summer estate on Nantucket. Recent work includes a state of the art music school built to LEED Gold standard, a penthouse apartment for a young family in Boston and a summer compound on the seashore. Design philosophy: to strike a harmonious balance between form and function.

设计理念：创建在形式与功能上的和谐。

Irina Markidonova, Ilona Menshakova. Sisters' Design, Moscow, Russia. Specialising in the design of private apartments, houses and gardens. Recent work includes an apartment for a contemporary art collector, a private spa complex at a country estate and a ground floor apartment with access to the surrounding landscaped gardens. Current projects include a loft apartment, a wooden house surrounded by pine forest and the company's office space in the historic centre of Moscow. Design philosophy: inspired by classic European tradition and its diverse range of cultural and architectural history.

设计理念：经典欧洲传统以及其各类文化与建筑历史中获悉的灵感。

Kamini Ezralow

Kamini Ezralow. Ezralow Design Ltd, London. A boutique design studio specialising in understated luxury residential and hospitality projects. Recent work includes a lateral apartment in the heart of Mayfair, a Swiss chalet and the refurbishment of guest rooms, suites and the outdoor space of the famed Marbella Club, inspired by the founders love of Californian living. Current projects include private residences in Dubai, a family house in London and a boutique hotel in Spain. Design philosophy: it's not just about environment, design from the inside.

设计理念：设计不单是外在环境，更多是来自其内在。

Fabio Galeazzo. Galeazzo Design, São Paulo, Brazil. A multidisciplinary team working in partnership with artisans to create projects that go beyond form and function. Recent work includes a colourful apartment overlooking the Panambi Park and several private houses. Current projects include the architectural and interior design concept for a fashion school including a museum of Brazilian fashion. Design philosophy: to create exclusive, luxury environments that are full of life.

设计理念：充实生活，创造专属、奢华的环境。

Fabio Galeazzo

Swank Interiors

Lanassir Lawes. Swank Interiors, Norfolk, UK. A residential and commercial interior design practise. Recent projects include an award winning restaurant in Norfolk, a New England style seafront home in Southwold and an apartment overlooking the marina in Ipswich. Current work includes a duplex penthouse apartment in Norwich, a grade II listed farmhouse in Norfolk and a barn conversion in rural Suffolk. Design philosophy: to enhance a building's architecture by fusing colour and texture.

设计理念：通过颜色与质地，来加强建筑物的建筑结构。

Beverley Williams. Beverley Williams Interiors, London, UK. Specialising in luxury residential interiors. Current projects include two penthouse apartments in the city of London and a 500 year old house in Oxfordshire. Recent work includes a 10,000 sq ft house in the West End of London, a family house in Mayfair and a farmhouse in Normandy. Design philosophy: clean lines and tailored spaces.

设计理念：简洁线条及专属空间。

Yu-shu Hsieh. Rainmark Design, Taiwan. Specializing in luxury residential and commercial interiors. Recent projects include coffee shops and restaurants in Taipei and several houses. Current work includes landscape and architectural design in Shanghai, a well known singer's holiday house plus furniture design. Design philosophy: innovative and passionate.

设计理念：革新与激情。

DESIGN INTERVENTION

Nikki Hunt & Andrea Savage. Design Intervention, Singapore. A multi award winning design consultancy who specialise in high end residential and boutique commercial projects. Current and recent work includes a hotel project in Niseko, Japan, a family home in Singapore and a beach house in Phuket, Thailand. Design philosophy: understanding the client is key.

设计理念：了解客户才是关键。

Hill House

Jenny Weiss & Helen Bygraves. Hill House Interiors, Surrey, UK. Specialising in both residential and commercial luxury interiors globally. Current projects include the refurbishment and interior of a duplex apartment in Belgravia, a contemporary beachside residence in Sandbanks, Poole and the specification and interior of a modern mansion in Surrey. Recent work includes New Oyster 885 luxury sailing boat, a modern mansion in St George's Hill, Weybridge, a Grade II Listed residence in Chester Square and a luxury 5 bedroom London riverside apartment. Design philosophy: thinking outside the box.

设计理念：跳出固有思维模式。

Elena Akimova & Ekaterina Andreeva

Elena Akimova & Ekaterina Andreeva, Moscow, Russia. Working on a variety of projects from galleries to private luxury apartments and villas in Russia, Spain, France and Austria. Recent work includes scenography for the ArtBasel Fair 2013, the interior concept for Gallery 9/11 in Moscow, several private apartments in Moscow, their own exhibition space and scenography for the VIP lounge zone at Moscow Design Week 2010, 2011 and 2012. Current projects include a medical centre, two apartments in Vienna plus three apartments in Moscow and a villa in Ibiza. Design philosophy: can't live without humour and imagination.

设计理念：生活不能没有幽默与想象力。

Cathy Kincaid

Cathy Kincaid. Cathy Kincaid Interiors Ltd, Dallas, Texas, USA. A leading designer who has transformed homes over the past 30 years. Recent projects include a family home and an historic cottage in Highland Park, Dallas and a Fifth Avenue pied-à-terre in New York. Current work includes a Spanish Colonial home, a Palladian Estate in Dallas and a Normandy style home in Buck County, PA. Design philosophy: traditional, comfortable and elegant.

设计理念：传统、舒适与高端。

Suna Interior Design

Helen Fewster & Rebecca Tucker. Suna Interior Design Ltd, London, UK. Suna Interior Design is an award-winning, London based, boutique interior design consultancy providing design services for property developers and the hospitality industry. Recent work includes the full design for a large development in Wandsworth, an exclusive apartment in Cambridge and the full interior specification design for 45 apartments in East London. Current

projects include the complete refurbishment of six apartments and four junior suites including interior and specification design for a boutique hotel in Mayfair, London, a four bedroom home in Cambridge and a bespoke pop-up gallery show house in Portobello, London. Design philosophy: contemporary and considered interior design.

设计理念：现代的、精心设计的室内设计。

Alexandra Schauer

Alexandra Schauer. Alischa Interior Design, Vienna, Austria. Work is international, specialising in custom made interiors, luxury chalets and private residences including hotels and castles. Recent work includes Hotel Stein in Salzburg, townhouses in Vienna, a chalet in Kitzbühel and a finca in Ibiza. Current projects include a series of residential projects in the mountains of Vienna, Salzburg and Tirol, the complete restoration of a villa in Munich, the renovation of a 400 sq m apartment in Moscow, an historic palace in Vienna, a lake house in Attersee and a private yacht in Italy. Design philosophy: harmony, luxury and comfort.

设计理念:和谐、奢华与舒适。

443

Christian Baumann. ABRAXAS Interieur, Zurich, Switzerland. Specialising in custom made, high end projects for an international clientele. Recent work includes interior design for a mountain apartment in Klosters, the complete interior design of the bar, meeting room and guestrooms in a boutique City Hotel in Winterthur and the concept

for two new apartments in one of the newest skyscrapers in the business district of Zurich. Current projects include a new build chalet in the Swiss mountains, a high end rooftop residence in downtown Zurich and a new build apartment in Lichtenstein. Design philosophy: designers are meant to be loved, not understood.

设计理念：设计师应被喜爱，而不是被理解。

Nikolay Tsupikov

Nikolay Tsupikov. Tsupikov Nikolay Interiors, Moscow, Russia. Established in 2003, the studio specialises in the complete design of luxury private residences. Recent work includes a retro/futuristic styled town house in the Moscow suburbs, current projects include several private residences in Russia and a neoclassical styled apartment in the centre of Moscow. Design philosophy: to find the balance between beauty and comfort.

设计理念：寻求美感与舒适之间的平衡。

Marylou Sobel

Marylou Sobel. Marylou Sobel Interior Design, Sydney, Australia. A boutique practice specialising in luxurious, residential design. Recent works include the refurbishment of an apartment at Circular Quay with spectacular views of the Harbour Bridge and Opera House, the renovation of an Italianate house in Centennial Park and a kitchen and conservatory in a home in Vaucluse. Current projects include the refurbishment of a waterfront home in Rose Bay, the renovation of a large covered terrace in Bellevue Hill and a house in Watsons Bay. Design philosophy: creating spaces that are pleasing to the heart and soul, in harmony with their surroundings.

设计理念： 空间的创建是为了使心与灵魂得到愉悦，使和谐伴随身旁。

Matrix Design

Guan Wang, Jianhui Liu, Zhaobao Wang. Shenzhen Matrix I.D. China. Founded in early 2010 with a team of young and prominent designers. Recent works includes Chengdu Vanke Office, Jin Yu Ti Xiang Vacation Clubs in Sichuan Province and Zhengzhou Vanke Project sales centre. Current projects include Wandao Project sales centre in Wandao Real Estate, Yuewan Villa Show Flat in Chongqing and Model House of Shan Yu Qing Hui in Shenzhen. Design philosophy: to create a new world.

设计理念：打造一个新世界。

Tineke Triggs. Artistic Designs for Living (ADL) San Francisco, California, USA. Specialising in luxury residential interior architecture and design. Current projects include a modern, high tech estate in Silicon Valley, a grand Victorian home in the exclusive Pacific Heights neighbourhood and a family retreat in the private community of Martis Camp, Lake Tahoe. Recent works includes a ranch in St. Helena, a private residence in Los Angeles and an artists retreat in Marin. Design philosophy: push the limits and break the rules to create unique spaces with enduring appeal.

设计理念：超越极限，打破常规才是创建独特空间经久不衰的吸引力。

463

GLAMOROUS

Yasumichi Morita. GLAMOROUS co. ltd, Tokyo, Japan. Established in 2000 the design office has since broadened its appeal to include projects in Hong Kong, Singapore, New York, London, Russia, Qatar and other major cities. Recent work includes B-one buffet restaurant at The Sherwood Taipei, Sunluxe Collection jewellery shop at One Central/MGM Macau and Galerie du Nord, bar in Osaka. Current projects include Morimoto South Beach Japanese restaurant at the Shelborne Wyndham Grand, Miami Beach, Isetan Shinjuku store renovation in Tokyo and a Baccarat chandelier to be displayed at Yebisu Garden Place, Tokyo in winter 2014. Design philosophy: glamour is key.

设计理念：魅力才是关键。

Designers in the City

Sonia Warner & Jacinta Woods. Designers in the City, NSW Australia. A boutique practice specialising in both residential & commercial projects. Current work includes a collection of contemporary apartments in

the heart of the city, along with a number of impressive family homes in Sydney's Northern & Eastern suburbs. Recent projects include the concept and project management of holiday houses on the famous coastline of Sydney's Palm Beach, whilst collaborating with a number of architects designing functional spaces in heritage listed properties. Design philosophy: to create unique spaces which exceed client expectation.

设计理念：超越客户期望，打造独一无二的空间。

Christine Klein

Christine Klein. CKlein Properties, California, USA. Specialising in cost effective, eco friendly property renovation. Recent work includes the creation of Hammer and Nails Salon, the first for men only, on Melrose Avenue in Los Angeles, a Marina Del Rey penthouse, a glamorous laundry and powder room and the update of a West Los Angeles condominium. Current projects include an LA townhouse and a mid century West Hollywood penthouse. Design philosophy: to redesign energy efficient living spaces.

设计理念：重新设计高效节能的生活空间。

4 Paolo Moschino and Philip Vergeylen
Nicholas Haslam Ltd
12-14 Holbein Place
London SW1W 8NL
Tel: +44 (0) 207 730 8623
Fax: +44 (0) 207 730 6679
design@nicholashaslam.com
www.nicholashaslam.com

12 Eric Kuster
Metropolitan Luxury
Sparrenlaan 11, 1272 RN Huizen
The Netherlands
Tel: +31 355 318 773
Fax: +31 355 318 770
info@erickuster.com
www.erickuster.com

18 Mary Douglas Drysdale
Mary Douglas Drysdale Inc
2026 R Street, NW
Washington DC 20009 USA
Tel: +1 202 588 0700
marydouglasdrysdale@yahoo.com
www.marydouglasdrysdale.com

26 James van der Velden
BRICKS Amsterdam
Nieuwezijds Voorburgwal 340
1012 RX Amsterdam
The Netherlands
Tel: +31 (0) 6 2120 1272
info@bricks-amsterdam.com
www.bricksamsterdam.com

30 Allison Paladino
Allison Paladino Interior Design
6671 W Indiantown Road
Ste 50-435 Jupiter, FL 33458, USA
Tel: +1 (561) 741 0165 (ext 2)
Fax: +1 (561) 741 0169
zita@apinteriors.com
www.apinteriors.com

34 Ming Xu & Virginie Moriette
Design MVW Co Ltd
No 322 Xingguo Road
Changming District
Shanghai, China
Tel: +86 21 628 155 94
contact@design-mvw.com
www.design-mvw.com

40 Helene Hennie
Christian's & Hennie
Skovveien 6, 0257 Oslo, Norway
Tel: +47 2212 1350
Fax: +47 2212 1351
info@christiansoghennie.no
www.christiansoghennie.no

46 Wan FuChen
Suzhou FuChen Design Studio
266 Loumen Road
Suzhou Industrial Park
Ping Jiang Design Room 7-102
Suzhou, China
Tel: +86 (512) 6935 3585
tmfcs@126.com
www.fcd888.com

50 Marc Hertrich & Nicolas Adnet
Studio Hertrich & Adnet
5 Passage Piver
75011 Paris, France
Tel: +33 (0) 1 43 14 00 00
Fax: +33 1 43 38 86 01
contact@studiomhna.com
www.studiomhna.com

56 Mr Yu Feng
Deve Build Interior Design Institution
619 Block B
Fortune Plaza
Bonded Area
Futian District, Shenzhen, China
Tel: +86 755 8212 9205
Fax: +86 755 8212 9205
szdayu@126.com
www.devebuild.com

60 Meryl Hare
Hare + Klein
Level 1, 91 Bourke St
Wooloomooloo, NSW
2011 Australia
Tel: +612 9368 1234
Fax: +612 9368 1020
info@hareklein.com.au
www.hareklein.com.au

66 Jorge Canete
Interior Design Philosophy
Chateau de St Saphorin sur Morges
Switzerland
Tel: +41 78 710 2534
Fax: +41 21 944 3757
info@jorgecanete.com
www.jorgecanete.com

72 Chang Ching-Ping
Tien Fun Interior Planning Co
12F, No 211 Chung Min Road
North District
Taichung City 404 Taiwan R.O.C.
Tel: +886 4 220 18908
Fax: +886 4 220 36910
tf@mail.tienfun.tw
www.tienfun.com.tw

80 Emma Sims Hilditch
Sims Hilditch
The White Hart
Cold Ashton
Gloucestershire, SN14 8JR
Tel: +44 (0) 1249 783 087
emma@simshilditch.com
www.simshilditch.com

84 Chen Yi & Zhang Muchen
Beijing Fenghemuchen Space Design
Centre, 4-B-1402 Bai Zi Wan
Hou Xian Dai Cheng
Chao Yang District
Beijing, China
Tel: +86 (10) 877 32 690
Fax: +86 (10) 877 65 176
muchenfenghe@163.com
www.muchenfenghe.com

90 Carl Emil Knox
Carl Emil Knox Design
Allschwilerweg 67
CH-4102 Binningen, Switzerland
Tel: +41 (0) 79 195 2308 & UK
Tel: +44 (0) 754 027 7700
info@carlemilknox.com
www.carlemilknox.com

94 Ben Wu
W. Architectural Design Co Ltd
6F No 1298 Middle Huaihai Road
Shanghai, China
Tel: +86 (21) 342 30218
Fax: +86 (21) 540 77476
wubin516@hotmail.com

100 Stefano Dorata
Studio Dorata
00197 Rome
12a/14 Via Antonio Bertoloni
Italy
Tel: +39 (0) 6808 4747
Fax: +39 (0) 6807 7695
studio@stefanodorata.com
www.stefanodorata.com

106 Yu-Lin Shin
Dumas Interior Design Group
10F - 10 No 161, Gongyi Road
West District
Taizhong City, Taiwan
Tel: + 886 423 050 585
Fax: +886 423 050 565
dumas@dumas-design.com
www.dumas-design.com

112 Ligia Casanova
Atelier Ligia Casanova
Rua das Pracas
30 - 3, 1200 - 767 Lisbon, Portugal
Tel: +351 919 704 583
atelier@ligiacasanova.com
www.ligiacasanova.com

118 Nicky Dobree
Nicky Dobree Interior Design Ltd
25 Lansdowne Gardens
London SW8 2EQ
Tel: +44 207 627 0469
info@nickydobree.com
www.nickydobree.com

124 Kunihide Oshinomi
k/o design studio
2-28-10 # 105 Jingumae
Shibuya-Ku Tokyo
Japan 150 - 0001
Tel: +81 3 5772 2391
Fax: +81 3 5772 2419
info@kodesign.co.jp
www.kodesign.co.jp

128 Aleksandra Laska
Aleksandra Laska
U.L. Krakowskie Przedmiescie 85 m5
00-079 Warsaw
Poland
Tel: +48 609 522 942
Fax: +48 228 260 796
aleksandra.laska@gmail.com

134 Maria Barros
Maria Barros Home, Av. De Sintra 533
2750-496 Cascais
Portugal
Tel: +351 21 485 2976
info@mariabarroshome.com
www.mariabarros.com

138 Jianguo Liang, Wenqi Cai, Yiqun
Wu, Junye Song, Zhenhua Luo
Chunkai Nie, Eryong Wang
Beijing Newsdays Architectural Design
Co Ltd, Jia 10th
Bei San Huan Zhong Road
West City District Beijing
P.R. China 100120
Tel: +86 10 820 86 969
Fax: +86 10 820 87 899
newsdays@newsdaysbj.com
www.beijingnewsdays.com

144 Christopher Dezille
Honky Architecture & Design
Unit 1 Pavement Studios
40-48 Bromells Road
London SW4 OBG
Tel +44 207 622 7144
Fax: +44 207 622 7155
chris@honky.co.uk
www.honky.co.uk

150 Charlotte & Poppy O'Neil
POCO Designs, 24 Glenmore Road
Paddington 2021, Sydney, Australia
Tel: +2 8356 9632
Fax: +2 9380 9486
poppy@pocodesigns.com.au
www.pocodesigns.com.au

156 Hong Zhongxuan
HHD Eastern Holiday International
Design Institution, Unit B01-B09
G/F Designer Centre
No 72 Longzhu Road North
Nanshan District, SZ China
Tel: +86 755 266 040 89
Fax: +86 755 266 039 77
hhdchina@hhd.hk

162 Angelos Angelopoulous
Angelos Angelopoulos Associates
5 Frinonos Str 11636, Mets
Athens, Greece
Tel/Fax: +30 210 756 7191
design@angelosangelopoulos.com
www.angelosangelopoulos.com

166 Idmen Liu
Shenzhen Juzhen Mingcui Design Co
Ltd, 2A Room No 6 Building
Bishuilongting, Boan District
Shenzhen P.R. China
Mobile: +86 (0) 159 866 95159
Tel: +86 (0) 186 803 61907
vivian_158@163.com
www.idmen.com.cn

172 Holly Stone & Nashmita Rajiah
Intarya, 8 Albion Riverside
8 Hester Road, London SW11 4AX
Tel +44 207 349 8020
info@intarya.com
www.intarya.com

176 Elin Fossland
ARKITEKTFOSSLAND AS
Kirkegata 8, 3016 Drammen, Norway
Mobile: 916 64 684
Tel: +47 32 847 454
elin@arkitektfossland.no
www.arkitektfossland.no

182 Carmo Aranha & Rosario Tello
SAARANHA&VASCONCELOS
Rua Vale Formoso, N 45
1950-279 Lisbon, Portugal
Tel: +351 218 453 070
Fax: +351 218 495 325
info@saaranhavasconcelos.pt
www.saaranhavasconcelos.pt

188 Gu Teng
Times Property Holdings Limited
Times Property Centre
410 Dongfeng Zhong Road
Guangzhou, China 510030
Tel: +86 20 834 86 668
Fax: +86 20 834 86 788
guteng@timesproperty.cn
www.timesgroup.cn

194 Toni Espuch
AZULTIERRA
C/Córcega 276-282
08008 Barcelona, Spain
Tel: +93 217 8356
maribel.lope@azultierra.es

198 Maurizio Pellizzoni
MPD London (Maurizio Pellizzoni
Design Ltd)
75-81 Burnaby Street
Fairbank Studios, Studio 8
London SW10 ONS
Tel: +44 207 352 3887
studio@mpdlondon.co.uk
www.mpdlondon.co.uk

204 Katharine Pooley
Katharine Pooley Ltd
160 Walton Street
London SW3 2JL
Tel: +44 (0) 207 584 3223
Fax: +44 (0) 207 584 5226
www.katharinepooley.com
& Katharine Pooley Doha
The Gate Mall, Maysaloun Street
West Bay, Doha, Qatar

210 Bronnie Masefau
Bronnie Masefau Design Consultancy
P.O. Box 8378 Armadale
VIC 3143 Australia
Tel: +61 402 040 703
bronnie@bronniemasefrauau.com.au
eileen@bronniemasefrauau.com.au
www.bronniemasefau.com.au

214 Ana Cordeiro
Prego Sem Estopa
Calçada do Combro 36
1200 - 114 Lisbon
Portugal
Tel: +351 213 421 583
Fax: +351 213 429 719
ana.crdr@gmail.com
www.pregosemestopa.pt

218 Prasetio Budhi
Plus Design
Jln. Bromo 10B, Guntur
Jakarta 12980, Indonesia
Tel: +62 21 8296 4751
Fax: +62 21 8379 6131
projects@plus-dsgn.com
www.plus-dsgn.com

222 Kari Arendsen & Jenn Bibay
Intimate Living Interiors
143 South Cedros Avenue
Suite C203, Solana Beach
CA 92075 USA
Tel: +1 858 436 7127
Fax: +1 858 436 7140
kari@intimatelivinginteriors.com
www.intimatelivinginteriors.com

226 Nick Candy
Candy & Candy Ltd
Rutland House
Rutland Gardens, London SW7 1BX
Tel: +44 207 590 1900
info@candyandcandy.com
www.candyandcandy.com

230 Laura Brucco
Laura Brucco
Catex 3228 PB C1425CD
Buenos Aires
Argentina
Tel: +54 11 4808 9565
estudio@laurabrucco.com
www.laurabrucco.com

234 Joseph Sy
Joseph Sy & Associates
17/Fl. Heng Shan Centre
141-145 Queen's Road East
Wan Chai
Hong Kong
Tel: +852 2866 1333
Fax: +852 2866 1222
design@jsahk.com
www.jsahk.com

240 Alejandro Niz & Patricio Chauvet
Niz + Chauvet Arquitectos
General Marcial Lazcano # 20
Col. San Angel
Del. Alvaro Obregon
C.P. 01060 Mexico City
Mexico
Tel: +52 55 56 16 1801
taller@niz-chauvet.com
www.niz-chauvet.com

246 Tim Campbell
Studio Tim Campbell
746 S. Los Angeles St. Suite 810
Los Angeles, CA 90014
& 19 Hudson Street
No. 501 New York 10013, USA/504
Tel: +1 213 688 1440
Fax: +1 213 688 1442
tanis@studiotimcampbell.com
www.studiotimcampbell.com

250 Ben & Hamish Lewis
TRENZSEATER
121 Blenheim Road
Christchurch, 8041 New Zealand
Tel: +64 3 343 0876
& 80 Parnell Road, Parnell
Auckland, 1052 New Zealand
Tel: +64 9 303 4151
benlewis@trenzseater.com
www.trenzseater.com

254 Olga Stupenko
Olga Stupenko Design
5 Bolshoy Chudov Lane
119270, Moscow, Russia
Tel: +7 499 246 6706
Fax: +7 495 773 6440
& Unit 12 Wheatley Business Centre
Old London Road, Wheatley
Oxfordshire OX33 1XW
Tel: +44 1865 875 504
info@olgastupenko.com
www.olgastupenko.com

258 Jiang Jianyu
Hangzhou Daxiang Art Design Co Ltd
Room 809, No 18 Xidoume Road
Xihu District, Hangzhou, China
Tel: +86 571 8755 5966
Fax: +86 571 8755 5955
dx87555966@163.com

264 Elaine McBride & Gillian Monro
Platform 9 International Ltd
Garfield House, Cumbernauld Road
Stepps, Glasgow, Scotland G33 6HW
Tel: + 44 141 779 9449
Fax: +44 141 779 9559
info@platform-9.com
www.platform-9.com

268 Stacey Sibley
Alexander James Interior Design
8 The Pavilions, Ruscombe Business
Park, Twyford, Berkshire RG10 9NN
Tel: +44 (0) 118 932 0828
sandra@aji.co.uk
www.aji.co.uk

272 Detlev Böhnke
Paris 56 - fine interiors
Pariser Str 56, 10719 Berlin, Germany
Tel: +49 30 345 06957
Fax: +49 30 345 06958
berlin@paris56.de
www.paris56.de

276 Suzanne Lovell
Suzanne Lovell Inc
225 West Ohio Street, Suite 200
Chicago, Illinois 60654 USA
Tel +1 312 595 1980
Fax: +1 312 595 9295
contact@suzannelovellinc.com
www.suzannelovellinc.com

282 Bo Li
Cimax Design Engineering Limited
7B Building 9, Haiying Chang Chen No 2
Wenxin Road, Shenzhen City
Guangdong Province, P.R. China
Tel: +86 0755 2644 8677
Fax: +86 0755 2644 8677
libodesign@126.com
www.libodesign.com

286 David Powell & Fenwick Bonnell
Powell & Bonnell, 236 Davenport Road
Toronto, Ontario, Canada M5R 1J6
Tel: +416 964 6210
Fax: +416 964 0406
info@powellandbonnell.com
www.powellandbonnell.com

292 Pippa Paton
Pippa Paton Design Ltd
West Wing, Old Berkshire Hunt
Kennels, Oxford Road, Kingston
Bagpuize, Oxon OX13 5AP
Tel: +44 (0) 1865 595470
Mobile: +44 (0) 7836 793 624
studio@pippapatondesign.co.uk
www.pippapatondesign.co.uk

298 Tomoko Ikegai
ikg inc
210 Building, 2-1-5-3F Shimomeguro
Meguro-ku, Tokyo 153-0064 Japan
Tel: +81 3 6417 4817
Fax: +81 3 6417 4816
info@ikg.cc
www.ikg.cc

304 Julia Buckingham
Buckingham Interiors + Design
1820 + 1822 W. Grand Avenue
Chicago, Il 60622 USA
Tel: +1 (312) 243 9975
Fax: +1 (312) 243 9978
info@buckinghamid.com
www.buckinghamid.com

308 Ann Yu
Guangzhou DOMANI Decoration
Design Co Ltd, No 275 Fanglun Road
Liuan District, Guang Zhou, China
Tel: +86 020 8180 6911
Fax: +86 020 8154 0266
domani_dcbiz@vip.163.com
www.domaniink.net

312 Karen Akers
Designed by Karen Akers
58 Middle Street
McMahons Point
NSW 2060 Australia
Tel: +61 2 892 09018
Fax: +61 4 341 51878
info@karenakers.com.au
www.karenakers.com.au

316 Kari Foster, Angie Pache
Jill Bosshart, Renee Keller
Rachel Blackburn, Michaela Jenkins
Jason Schleisman
Associates III Interior Design
1516 Blake Street
Denver, Co 80202, USA
Tel: +1 303 534 4444
Fax: +1 303 629 5035
amy@associates3.com
www.associates3.com

320 Shuheng Huang
Sherwood Design
Room 2, 17F No 2, Lane 150, Sec 5
Sinyi Road, Sinyi District
Taipei City 110, Taiwan
Tel: +886 02 6636 5788
Fax: +886 02 6636 5868
sh@sherwood-inc.com
www.sherwood-inc.com

326 Jennifer Mehditash
Mehditash Design
821 Bellis Newport Beach
Calforina -2005 Palmer Ave
#285 Larchmont
New York 10538, USA
info@mehditashdesign.com
www.mehditashdesign.com

330 Karen Howes
Taylor Howes Designs
29 Fernshaw Road
London SW10 OTG
Tel: +44 207 349 9017
admin@taylorhowes.co.uk
www.taylorhowes.co.uk

336 Gloria Cortina
Gloria Cortina Estudio
Montañas Rocallosas
505 Lomas Virreyes C.P. 11000
México, D.F.
Tel: +52 (55) 552 054 63
Fax: +52 (55) 520 294 22
equipo@gloriacortina.mx
www.gloriacortina.mx

340 Steve Leung
Steve Leung Designers Ltd
30/F Manhattan Place
23 Wang Tai Road, Kowloon Bay
Hong Kong, China
Tel: +852 2527 1600
Fax: +852 3549 8398
sld@steveleung.com
www.steveleung.com

344 Patrick Tyberghein
Carter Tyberghein
1 Melbray Mews
Hurlingham Road
London SW6 3NS
Tel: +44 (0) 207 731 6557
Fax: +44 (0) 207 731 6179
info@cartertyberghein.com
www.cartertyberghein.com

348 Ajax Law & Virginia Lung
One Plus Partnership Ltd
Unit 1604, 16th Floor
Eastern Centre
1065 King's Road
Hong Kong, China
Tel: +852 2591 9308
Fax: +852 2116 1392
admin@onepluspartnership.com
www.onepluspartnership.com

356 Mona Hajj
Mona Hajj Interiors
13 East Eager St. Baltimore
MD 21202
USA
Tel: +1 410 234 0091
Fax: +1 410 234 0098
mh@monahajj.com
www.monahajj.com

360 Oleg Klodt & Anna Agapova
Architecture & Design by Oleg Klodt
B. Sukharevskiy side street 11/1
Moscow
Russia 127051
Tel: +7 (495) 221 11 58
Fax: +7 (495) 221 11 58
info@olegklodt.com
www.olegklodt.com

364 Jin Jian
ZPSS (Zhu Ping Shang She)
Floor 4
Xinshidai no 808 Gudun Road
Hangzhou, China
Tel: +86 (571) 8894 9785
834237702@qq.com

370 Broosk Saib
Broosk Saib
Flat 1, 36 Wilton Crescent
London SW1X 8RX
Tel +44 (0) 207 235 0787
broosk@broosk.com
www.broosk.com

376 Sheena Notley-Griffiths, Claudia
Fuchsberger, Adriel Lack, Henry
Prideaux
The Studio, Harrods, 3rd Floor
87-135 Brompton Road
Knightsbridge
London SW1X 7XL
Tel: +44 207 225 5926 ext: 5622
www.thestudioatharrods.com

380 Kathleen Hay
Kathleen Hay Designs
PO Box 801, 8 Williams Lane
Nantucket
MA 02554, USA
Tel: +1 508 228 1219
Fax: +1 508 228 6366
kathleenhay@comcast.net
www.kathleenhaydesigns.com

386 Irina Markidonova & Ilona
Menshakova
Sister's Design
Molochny per. 11-5
119034 Moscow, Russia
Tel: +7 985 999 5084
Fax: +7 926 262 82 22
sisters-design@mail.ru
www.sisters-design.ru

392 Kamini Ezralow
Ezralow Design Ltd
Studio A5, 1927 Building
The Old Gasworks, Michael Road
London SW6 2AD
Mobile: +44 (0) 773 495 6988
Tel: +44 (0) 207 736 9668
kamini@ezralowdesign.com
www.ezralowdesign.com

398 Fabio Galeazzo
Galeazzo Design
Rua Antônio Bicudo
83 - Sao Paulo/SP, Brazil
Tel: +55 (11) 3064 5306
info@fabiogaleazzo.com.br
www.fabiogaleazzo.com.br

402 Lanassir Lawes
Swank Interiors
Three Gates Farm
Fen Street, Bressingham
Norfolk IP22 2AQ
Tel: +44 (0) 1379 687 542
lanassir@swankinteriors.co.uk
www.swankinteriors.co.uk

406 Beverly Williams
Beverley Williams Interiors
78 York St
London W1H 1DP
Mobile: +7887 576 069
Tel: +44 207 138 2826
info@beverley-williams.com
www.beverley-williams.com

410 Yu-shu Hsieh
Rainmark Design
2F No 399, Fujin St, Taipei City
Songshan Dist. 10583, Taiwan
Tel: +886 2 3765 3556
Fax: +886 2 2767 5205
rainmark.design@gmail.com
www.rainmark-design.com

414 Nikki Hunt & Andrea Savage
Design Intervention I.D.
75E Loewen Road, Tanglin Village
Singapore 248845
Tel: +65 9623 7593
Fax: +65 6468 7418
info@diid.sg
www.designintervention.com.sg

418 Jenny Weiss & Helen Bygraves.
Hill House Interiors
32-34 Baker Street
Weybridge
Surrey KT13 8AU
Tel: +44 (0) 1932 858 900
Fax: +44 (0) 1932 858 997
design@hillhouseinteriors.com
www.hillhouseinteriors.com

422 Elena Akimova & Ekaterina
Andreeva
119034, Russia
Moscow
2nd Zachatievskiy Lane 11, ap.16
Tel: +7 (925) 518 4960
& +43 (676) 942 9210
elena1218akimova@gmail.com
& 5261408@gmail.com

428 Cathy Kincaid
Cathy Kincaid Interiors Ltd
4007 Normandy
Dallas, Texas 75205 USA
Tel: +1 214 522 0856
Fax: +1 214 528 3527
info@cathykincaidinteriors.com
www.cathy-kincaid.com

434 Helen Fewster & Rebecca Tucker
Suna Interior Design Ltd
First Floor, 12a Deer Park Road
London SW19 3UQ
Tel: +44 208 544 9350
lucinda@sunainteriordesign.com
www.sunainteriordesign.com

438 Alexandra Schauer
Alischa Interior Design
Sternwartestrasse 73
1180 Vienna, Austria
Tel: +43 676 407 1407
Fax: +43 1 369 26 31
ali.schauer@a1.net
www.alischa.at

442 Christian Baumann
ABRAXAS Interieur GmbH
Hegibachstrasse 112
8032 Zurich, Switzerland
Tel: +41 44 392 21 92
Fax: +41 44 392 21 93
info@abraxas-interieur.ch
www.abraxas-interieur.ch

446 Nikolay Tsupikov
Tsupikov Nikolay Interiors
Moscow, Russia
Tel: +7 903 960 8633
nikolay@tsupikov.ru
www.tsupikov.ru

450 Marylou Sobel
Marylou Sobel Interior Design
No 1 Boronia Road
Bellevue Hill
2023 Sydney NSW, Australia
Tel: +61 2 9130 5899
Mobile: +61 411 195 404
Fax: +61 2 9130 1018
info@marylousobel.com.au
www.marylousobel.com.au

456 Matrix Team
Shenzhen Matrix Interior Design Co Ltd
Room 1805, Excellence Building
Fuhua 1 Road, Futian District
Shenzhen, P.R. China
Tel: +86 (0) 755 832 22578
Fax: +86 (0) 755 832 28692
wzb@matrixdesign.cn
www.matrixdesign.cn

460 Tineke Triggs
Artistic Designs for Living
2152 Union Street
San Francisco, CA, USA
Tel: +1 415 567 0602
Fax: +1 415 567 0604
tineke@adlsf.com
www.adlsf.com

466 Yasumichi Morita
GLAMOROUS co. ltd
2F 2-7-25 Motoazabu
Minato-ku
Tokyo 106-0046 Japan
Tel: +81 3 5475 1037
Fax: +81 3 5475 1038
info@glamorous.co.jp
www.glamorous.co.jp

470 Sonia Warner & Jacinta Woods
Designers in the City
330B Miller Street
Cammeray, NSW 2062
Australia
Tel: +02 995 44901
Fax@ +02 995 44901
jacinta@designersinthecity.com.au

474 Christine Klein
CKlein Properties
9000 W. Sunset Blvd
Suite 1100, W. Hollywood
CA 90069, USA
Tel: +1 310 623 1380
christine@ckleinproperties.com
www.ckleinproperties.com

图书在版编目（CIP）数据

第18届安德鲁·马丁国际室内设计大奖获奖作品：英文／（英）沃勒编著．－－南京：江苏凤凰科学技术出版社，2014.10
ISBN 978-7-5537-3752-2

Ⅰ．①安 Ⅱ．①沃 Ⅲ．①室内装饰设计－作品集－世界－现代－英文 Ⅳ．①TU238

中国版本图书馆CIP数据核字(2014)第203230号

© 2014 Andrew Martin International

第18届安德鲁·马丁国际室内设计大奖获奖作品

编　　　著	（英）马丁·沃勒
责 任 编 辑	刘屹立
特 约 编 辑	杜玉华

出 版 发 行	凤凰出版传媒股份有限公司
	江苏凤凰科学技术出版社
出版社地址	南京市湖南路1号A楼，邮编：210009
出版社网址	http://www.pspress.cn
总 经 销	天津凤凰空间文化传媒有限公司
总经销网址	http://www.ifengspace.cn
经 销	全国新华书店
印 刷	深圳市汇亿丰印刷科技有限公司

开 本	965 mm×1 270mm　1／16
印 张	30
字 数	24 000
版 次	2014年10月第1版
印 次	2015年6月第2次印刷

标 准 书 号	ISBN 978-7-5537-3752-2
定 价	598.00元（精）

图书如有印装质量问题，可随时向销售部调换（电话：022-87893668）。